Roswell Rediscovered

Steven Markus

Lighthouse Books for Translation and Publishing

Roswell Rediscovered by Steven Markus

Email: info@lighthousebooks.org

ISBN 13: 9783598131196

ISBN 10: 3598131194

Roswell Rediscovered

New and updated insight into the mysterious craft that crashed near Roswell, New Mexico in July 1947. Standard Roswell UFO lore sorted fact from fiction by Anna Jones and interviews with The Wanderling.

"Swooping in behind and over his left shoulder from out of the northwest and only a few hundred feet above the top of the truck were 2 large sharp-edged almost flat circular-shaped objects, blunt across the back and seemingly made of metal. The objects were flying side-by-side with one slightly in front, both headed ESE out over the horizon at an incredible high rate of speed. In only the few seconds it takes the boy to scramble up from under the dash, the objects are gone, leaving in their wake only a small residue lingering in the air like the smell of electricity and a quarter-mile wide swath of thick swirling air laying turbulently above the treetops like a sweltering mirage over a desert dry lake." (source)

In the summer of 1947 a young boy was traveling with his uncle visiting various archaeological sites and places of interest throughout the desert southwest. On Fourth of July weekend of that year, his uncle -- a notorious bio-searcher with strong ties to southwestern Native American cultures and who because of his discoveries will eventually have several plant species named after him -- had taken the boy as part of their exploration to learn firsthand about "The Long Walk" endured by the Navajos and Apaches as well as visit the gravesite of Billy the Kid near Fort Sumner, New Mexico.

As they were about to turn onto a main highway from a side road near Fort Sumner, they were stopped by a 5-or-6 truck convoy of military vehicles headed northeast at a high rate of speed. Several were carrying huge wooden crates, some covered with tarps. A few days later, a famous meteorite hunter informs the boy's uncle some mysterious objects have been found in the desert near Roswell that have an unknown writing on them. The uncle goes to Roswell to investigate and takes the boy with him. The following contains some of the previously unpublished insights taken from the boy's reminisces and conversations with his Uncle, interjected into, thus modifying standard Roswell UFO fare.

"With the lid of the box open, the foil mysteriously unfolds itself out of the container to about the size of a small handkerchief completely covering the box, the paper-thin foil displaying no sign of folds, creases, or wrinkles." (source)

"Retracing several miles in both directions of the suspected trajectory in an effort to confirm their conclusions, they discover a previously unknown and unspoiled touchdown point 5 miles from the debris field where the sand has somehow been crystallized. The plants and scrub brush growing along the periphery of the glass-like sand and gravel are not so much burnt or scorched as they are more-or-less trying to return to a natural growth stage after being severely wilted, apparently from whatever crystallized the sand 2 months earlier." (source)

PRELUDE TO THE CRASH

Tuesday, July 1, 1947:

Radar operators in central New Mexico using the then available WWII and mid 1940s rudimentary radar equipment at Roswell and Alamogordo and somewhat newer and much more sophisticated (and at the time top-secret) equipment used to track V-2 rocket launches at White Sands, unbeknownst to each other while on routine duty begin tracking an object that defies convention. The radar scopes project an image that appears to exhibit non-aircraft like maneuvering such as reversing direction in mid-flight, extreme right angle turns, able to climb straight up or down at ultra-high speeds, and the ability to stop immediately and hover in place.

Wednesday, July 2, 1947:

Mr. and Mrs. Dan Wilmot were sitting on the front porch of their home in Roswell, New Mexico about 10 o'clock in the evening when they observe a large circular object "like 2 inverted saucers faced mouth to mouth" zoom out of the southeast going in a northwesterly direction at a high rate of speed. The object disappears out of view over the treetops in the direction of the old limestone quarry on Six Mile Hill west of town.

Thursday, July 3, 1947:

Because of the unusual nature that the object exhibits on the radar screens, Roswell Army Air Field (RAAF) headquarters sends a technician to the White Sands Proving Ground (WSPG) in an effort to confirm consistency in readings between sites and locations. Lack of the correct clearance denies the technician total access to high levels of the WSPG radar grid. But what he does observe (and overhear among proving ground staff) is that their readings are similar to that of RAAF readings.

The object continues to go in-and-out of the WSPG radar detection system several times during a 24 hour period and in each instance (as had been observed on the previous occasions) always appearing to defy convention by exhibiting high speeds and non aircraft-like maneuvering. Checks between sites continue to substantiate no malfunction that would cause a similar return echo or radar imaging if the object being tracked was not "real". However, no visual sightings occur, nor are any reported by base or ground personnel.

With no change over 24 hours and the fact that there are NO inconsistencies, the technician is ordered back to Roswell. It is presumed the WSPG radar array continues tracking the object.

In 1947, the radar equipment used at White Sands was said to be composed of SCR-584 mobile units "modified and of an experimental nature" (read: secret) built into a K-78 trailer with a gross weight is 10 tons. The overall length 19.5 feet, width 8 feet, height 10 feet, 4 inches. Although at the proving grounds they were not intended to be moved, the unit was well capable of being moved or relocated quickly if special circumstances required it. As to the power, capabilities and efficiency of the SRC-584, on Oct. 14, 1947, it should be noted a SRC-584 was used to track Chuck Yeager in his record breaking supersonic flight of the Bell X-1 over Muroc Dry Lake as he accelerated to a speed of Mach 1.06 at an altitude of 42,000 feet.

As to the question if radar equipment was available and being used at RAAF on the nights in question, it is not clear. Official memos state that there was a lack of ground radar equipment convenient to Eighth Air Force Bases which are located at Fort Worth, Texas, Tucson, Arizona, and Roswell, New Mexico. However, there is a big difference between "lack of" "convenient" and none at all which opens the door to the possibility of radar capability not only at RAAF but from such sources as at the White Sands Proving Grounds.

To the Brass, lack of ground radar equipment "convenient to Eighth Air Force" means in relation to the implementation and accomplishment of the mandated Eighth Air Force mission which was bombing and for the most part at the time did not encompass superfluous saucer or bogie chasing except in how it might be related to the mission (say in a threat of attack, for example).

It has been stated that even if RAAF did have radar equipment available, it would not have the range to track objects at the distance of the impact site anyway. However for the most part, military electronic equipment is notoriously rated at the minimum distance to be used under the most extreme circumstances. Some transmitting devices stated as having a maximum of 350 mile range can operate 1200 miles (over 3 times the rated distance).

Radar devices are not much different. Plus enlisted men operators are constantly "hot rodding", jacking-up, or tweeking their equipment to maximize efficiency, especially when you can get away with it at small installations such as Roswell and especially so under circumstances that present themselves similar to the Roswell incident. Most of it can easily be chalked up as military double-speak for top-secret or classified equipment they don't want to talk about. Since the use of radar is an important early stage regarding the timeline of the Roswell incident, throwing doubt on its use or undermining any credibility of its use would be of the utmost priority.

Not known at the time because of the then in-place secrecy blanket but eventually revealed through obscure sources such as the footnotes in the book ICBM by G. Harry Stine; Orion Books, New York (1991), it is written that for the White Sands Missile Range V-2 testing program in the 1940s there were cameras, radars (including SCR-584 Doppler radio positioning system), and tracking telescopes linked with 100,000 miles of open wires run by the US Army Signal Corps. Not included in that on-base array were many interconnected "beyond the fence" remote sites.

For example, in 1947 there was a radar site for far-field tracking of missiles launched from White Sands located just north of U.S. 60 about 45 miles west of Socorro, New Mexico (there is nothing but a big gravel pit used by the state Department of Highways there now). Military presence in Catron County during the late 1940s was well known by most in the area, with the military staffing a second site the locals said was a radar tower on the road to the spread of rancher Marvin Ake 10 miles south of State Road 60 between Magdalena and Datil.

However, for the most part except for a previously "secret" radar site on the Proving Grounds proper near Oscura Peak in the San Andres Mountains that has since been revealed, no one has come forward or admitted that such sites operated or existed in other places. No doubt there were similar sites and locations on the east side of the proving grounds toward Corona and Roswell as well. Most likely, sites were placed outside the whole Proving Grounds perimeter in a close or overlapping set of intervals. If you draw a circle around White Sands with the launch site as the center and run the line through the site outside of Socorro, the circumference of that line runs almost right through Corona and toward Roswell and the Capitan Mountains. Such an interconnected array could have come into play relative to tracking an object similar to the Roswell object and fed into the Proving Grounds radar net control station but would not be able to be cited because of the secrecy. It seems highly unlikely they would have just the one site, possibly two, only in the vicinity of Socorro and not have them connected in some sort of much larger series.

Interestingly enough, less than a week before the Roswell Incident on May 29, 1947, a modified V-2 rocket called a Hermes B-1 vehicle (which was a highly classified top-secret project at the time) was launched that somehow inadvertently got wires physically crossed in the guidance system. Instead of heading up range toward the north as intended, the rocket headed south slamming into the Tepeyac Cemetery across the border in Mexico a mile and a half outside the city of Ciudad Juarez blasting out a crater 25 feet deep and 30 feet in diameter.

The tracking and recovery teams were prepared and in place for a crash-down at a pre-ordained impact site up range. However, the rocket traveling well over the speed of sound during its flight was miles off course and miles and miles away from the designated impact point. Even so, the U.S. Army was able to track it with investigators arriving on the scene, including travel time within 120 minutes. 2 weeks before a Hermes B-1 crashed out in the desert east of the impact zone on the outskirts of Alamagordo, New Mexico. Although the two Hermes B-1 vehicles that went awry were eventually put "on the record" (because information got out on them for no other reason than they slammed into semi-populated areas where they were seen), there were at least 4 others (and probably even more, flown from Complex 33) of which are not noted in contemporary records because of being so deeply classified at the time. Where they crashed has never filtered down into the public domain.

On Thursday, July 3rd -- the same day the RAAF radar technician was at White Sands -- the whole V-2 radar array is suspected of being on and in full up-and-running status because of a planned launch. When the brief and sketchy records that still exist regarding the July 3rd launch are traced, they indicate the launch was aborted because of an explosion on the pad, possibly claiming the deaths of two in a fire.

However, some sources indicate regardless of any fire or attempts to stop the rocket or what the records say or don't say, the failsafe mechanism did not work and the thing actually lifted off. Authorities are reluctant to let the public know the vehicle impacted into the desert west of the launch site on the Plains of San Augustin near Magdelena, New Mexico because of the Alamagordo incident followed by the Ciudad Juarez, Mexico incident that happened only a few days prior (which also crashed after an unresponsive shut-down mechanism failed to operate), hence the "explosion on the pad" story. In either case, the rocket is accounted for.

The secret V-2 radar array system -- which typically would NOT be in a full operational mode if there was no launch scheduled -- continues up and running in full swing because of the unrelated bogie that has shown up during the scheduled launch tracking sweep. What is happening here is IF the July 3rd launch never was -- that is if it never happened (and the fact remains that for the most part the launch has been erased, removed, or de-listed from a majority of the records) -- then the radar system would not be 'on'. If it wasn't 'on', then the object that eventually ends up being considered the Roswell UFO could not be tracked as successfully as it has been reported to have been tracked by using only the radar sites available at RAAF (see).

So did in fact the July 3rd launch take place? If the rocket blew up on the pad or took off wouldn't matter as the radar system would still be 'on'. However, some years later Grady L. "Barney" Barnett -- a soil conservation engineer for the Federal Government -- tells his friends L.W. Maltais and J.F. Danley who eventually come forward and maintained that on July 3, 1947, Barnett came across what he claims was a crashed "disk" in the course of his work on what he called "the Flats". "The Flats" are west of Socorro, New Mexico on the Plains of San Augustin not far from the Continental Divide. The location of the downed object was said to be between Datil and Horse Springs.

Barnett's description of what he found is that it was metallic and disc-shaped with a diameter of about 20-to-30 feet. The same diameter of the crater caused by the V-2 at Ciudad Juarez. The object that Barnett is said to have found is most likely not a disc but instead thought to be (if anything) the errant V-2 launched that same day pancaked onto the desert floor.

Ranch owners Jack Bruton and his wife heard whatever it was come down and went to investigate, locating the crash site. They found what they describe as a heap of twisted metal. Bruton did not know what it was but felt it was a piece of an airplane. There were no other parts nearby that he could be certain of. Bruton did not make public any information regarding numbers or markings anywhere on the rubble, however, because he felt it was a "piece of an airplane". He did report there was no sign of a pilot, crew, or passengers of any sort.

There is some talk of a diary kept by Barnett's wife that some people seem to see as undermining Barnett finding or seeing anything (disc or otherwise) during any of the particular days and dates in question. She notes in her diary that as far as she knew, her husband spent July 3-7, 1947 at his office in Magdalena.

The crash site is less than 60 miles from Magdalena, all basically wide open flatland desert roads. Supporters say if Barnett went from his home in Socorro to his office in Magdalena and then out to the Plains and came back, the wording of the diary entry for July 3rd period would allow that even though it doesn't say so specifically. The problem is that the diary makes no mention of anything "unusual" during or following that period.

It has been reported that while Barnett was at the crash scene, military personnel showed up and in a collaboration of testimony of sorts, one Marvin Ake (mentioned previously, a nearby rancher) reported seeing the military removing 2 truckloads of debris from the same general area. If such was the case, members of the military may have told Barnett not to mention what he saw to anyone including his spouse. Barnett, Bruton, or Ake who all report being at the crash scene relatively soon, make no reference to each other in any early accounts as well. Why Barnett -- a soil conservation engineer for the Federal Government -- felt there was so much work in the area of soil conservation that he was compelled to "work in his office" over the long Fourth of July weekend (a major Federal holiday) is not known.

There are reports that have Barnett and some students several miles south of the crash site at the then recently discovered Bat Cave searching for Native American artifacts. He may have been a guide although neither the diary nor comments by himself or others reflect any such scenario.

Barnett and the students are apparently aware of the crash somehow, either having seen something come out of the sky or having heard it. But they are unable to, some have said, locate the impact site because of nightfall. If such was the case, that would put Barnett at the crash scene a day after the impact. That is, early Friday morning July 4th rather than the afternoon of Thursday July 3rd. More than likely Barnett and the students were on the scene in the afternoon of the 3rd.

It is reported Barnett says while he was there, members of the military showed up. The confusion arises because other reports have it that the next morning (July 4th), the Army arrived at the crash scene with several trucks. "Arriving with several trucks the next morning" does not necessarily negate the possibility of a military presence of some sort arriving the day before. After all, Bruton and his wife had no problem getting to the scene on the 3rd. Plus if you recall, military investigators had no trouble reaching the scene (including travel time) within 120 minutes of the May 29th Ciudad Juarez crash which was not only miles and miles away from the designated impact point but also went in the totally wrong direction.

In an interesting twist of fate, the Very Large Array -- one of the world's premier astronomical radio observatories designed exclusively to probe deep space and consisting of 27 huge radio antennas on railroad tracks (dedicated in 1980) -- is located on the Plains of San Agustin 50 miles west of Socorro, not far from the Barnett site and within minutes of the location where the remote radar site (mentioned above) used to stand.

SCR-584 Mobile Radar Unit

The Crash

Friday, July 4, 1947:

Two Catholic nuns (reported to be Mother Superior Mary Bernadette and a Sister Capistrano) report seeing a bright fiery object appear to go to the ground well to the west and slightly north of Roswell, possibly in the mountains or beyond, late in the evening while looking out a third floor window of the now demolished Saint Mary's Hospital during the change of their shift, recording its passage in their logbook. They make no mention of an explosion, perhaps because of the distance the object went down from where they were (although some reports say the nuns saw a large flash in the night sky in the exact place on the horizon at the same time they lost eye contact with the object).

Years later, Dan Wilmot (mentioned above as sitting on the porch with his wife) came forward and said he too had seen a flash in the sky in the same direction around the same time that could have been an explosion. He stated he had been reluctant to come forward because the object he saw a few nights earlier (Wednesday, July 2nd) from the same porch location was a "flying saucer".

There is also some dispute as to the direction of the craft. Wilmot was initally quoted as saying the object zoomed in out of the southeast, going in a northwesterly direction. In his "years later" description the object comes in on a huge curve from the northwest crossing the local meridian only to continue its curve in a huge arc to the southwest. The dispute can easily be rectified if Wilmot is talking about 2 different nights and observations, which appears to be that he is.

At 11:27 P.M:

Because of the range of the non-secret 1947 radar equipment at individual sites was not comprehensive, all sites under a coordinated effort are ordered to give the object the utmost and intense scrutiny. At 11:27 P.M., one of the sites (secret or otherwise) in clear radar contact with the object reports it appears to divide, separate or come apart, then disappears off the screen. The belief is that the object crashed.

At about the same time, an archaeologist professor and some of his students taking advantage of a 3-day weekend are on site studies somewhere west of Roswell along the lower north slope of the Capitan Mountains looking for Native American artifacts and signs of pre-contact occupation when they observe an object fall to the ground not far from their camp.

Because of the darkness and impending storm, they make the decision to try to both locate and find out what it was the following morning. Sure that the object was not a meteor but an aircraft that crashed, one of the students is dispatched to find the closest phone. Sometime after midnight, the student circling around toward the small town of Capitan on Route 380 locates a phone and calls Sheriff George A. Wilcox in Roswell to inform him that they witnessed a plane crash of some type.

Wilcox calls the fire department (or volunteer fire department as the case may be because some reports say that at the time of the crash, 1947, Roswell did not have a full time paid fire department) and alerts them of the crash. Thinking there may be injured people, one fire truck with Dan Dwyer among others on it responds to the call escorted by members of the Roswell Police Department, making the rather long run west northwest to the impact area. The site is about 55 driving miles from Roswell near the edge of the Capitan Mountains, a few miles south of present day State Highway 246 (then SR-48) on the rough dirt road leading to the old Pine Lodge.

They arrive in the vicinity of the crash scene sometime well before the early morning hours and (it has been said by some) rather than take the truck through an unfamiliar area in what is left of the darkness, park not far from the main highway to wait for the Sun to come up. The question is why rush all the way out to the Pine Lodge area to locate a potential crashed plane with possible injured people then stop just after arrival because of being unfamiliar with the road?

Probably what happened is the Military had only just arrived on the scene themselves. And when the fire truck and police car contingent turned onto the dirt road from the main highway in the pre-dawn darkness, they were stopped by either MPs or an armed military patrol of some sort before they got very far. They were then placed not under actual military arrest per se but more-or-less in a quasi-holding pattern in an effort to maintain some sense of congeniality between the MPs and the two civilian firefighting and law enforcement agencies. The military authorities at that time (and especially so the outlying guards) were not fully aware of what the status of the situation is or even perhaps if the fire crew might eventually be needed or not.

It should be noted in regards to the Roswell fire department, there is a difference in opinion as to if the department even actually made runs outside the city at the time of the crash. There is a long and mutual history of any and all units assisting each other in the general area if the need arose. In 1947 as well as now, the idea was to help those in need and not fight amongst themselves.

According to department records, on June 21, 1947 Pumper No. 4 traveled 31 miles on a run "outside the city limit", the closest outside the city run before the crash. However, truth be said, NO runs were recorded or logged anywhere between July 2 and 6 outside the city. At the time, however, in 1947 (since it has been said the department was an all volunteer unit), record keeping may have been spotty at best. Plus the run did take off after midnight. Could be when everybody returned late the next day everybody thought they would just get caught up later and never did. It could also be chalked up to an example of a "requested" lapse of accurate record keeping on those specific dates from outside sources.

The Crash Site Stumbled Upon By Archaeologists, Capitan Mountains

(directions)

Saturday, July 5, 1947:

The Military, needing only to decipher the incident from the tracking of a single radar site, confirms the most recent data received. Then simply by using the last azimuth and the slant range from the radar readings, they immediately think they have determined the exact location and place of the downed craft. Because the object exhibited an ultra-high speed and unusual maneuvering and not being sure what it means, the Military moves in with a carefully selected team for investigation and/or possible recovery of the downed craft. However because the full sweep of the radar was blocked partially by a portion of the mountains giving an incomplete reading, the recovery team accompanied by armed soldiers are slowed locating the site.

At first light, the archaeologist Professor William Curry Holden of Texas Tech with the students who had been working sites with him in the area the previous day break camp to look for whatever they saw fall from the sky the night prior. Hiking in the general direction the object went down they stumble across the impact site, the object nearly sideways and fully positioned against the rocks. Later reports described it as looking like a crashed airplane without wings with a flat fuselage. Some reports imply the fuselage has a delta or wedge shape to it while others mention an almost circular crescent moon shape. All agree it was made of metal of some type.

In later interviews, Dan Dwyer is quoted as saying that he saw "the first pink lines of sunlight over the horizon" indicating being there at least pre-dawn of the morning of July 5th. He also notices an extremely strong glow showing up (not from a fire but similar to how lights illuminate the dark sky of a nighttime high school football game) over the crest of the hills away from the sunrise.

After sunup, Dwyer is able to sneak away undetected from the loosely watched or guarded fire crew and police officers, possibly by a planned or accidental diversion created by his buddies sharing hot coffee from thermos bottles with members of the military. He climbs up through the rocks, trees, and underbrush to a point where he is able to see a sizable number of uniformed military personnel, a series of now turned off floodlights, and various pieces of equipment such as jeeps, SRC-399 radio rigs, and other communication vans.

He sees as well the center of all the activity and what he describes later as a "strange craft" being lifted into the air by a crane and set on a flatbed truck. He continues to watch as it is secured with chains and cables then covered by a tarp. Because none of the fire fighters or police officers chose to join him and they remain basically under guard, no one of the group other than Dwyer is an actual eyewitness to the event.

The military recovery team, working most of the night under high-powered floodlights since sometime before 2:00 AM -- are just past the early stages of their investigation and completion of the setup of equipment around sunrise. Just as the sun is coming up but before Dwyer's unknown and totally undetected observation of a much more advanced stage of recovery an hour-or-two later, civilians suddenly show up on the scene. The soldiers are told by the team to escort them out of the area while other soldiers are ordered to search and secure a much wider perimeter and not just the dirt roads leading into the site.

The civilians mentioned are more than likely Holden and his students who had hiked in over the hills unexpectantly catching the military off guard. The Roswell fire and police department personnel (which technically might not be viewed as civilians by the public but considered as such by the military) are continued to be held or stopped by MPs some distance away along the dirt road leading up to the crash site.

Dwyer is able to slip back down the hill through the widening military sweep and rejoins his group undetected. Because Holden and his students arrived right on top of the actual crash scene much earlier than Dwyer's observation of the scene from a distance, neither Dwyer nor Holden or any member of either group is aware of the others presence.

Why the 2 groups do not eventually see each other in the "escort out" process is not fully known. Most likely, Holden's group left some equipment or at least cars or vehicles at or near their campsite. They are escorted back through the forest the same way they came rather than take them to the main highway down Pine Lodge Road past where the fire fighters and police officers are being held. After they drive from the campsite to the main road Holden and his students are told to park their vehicles and ordered to get out.

Since Holden's group is the only known unauthorized civilians that have actually seen and know about the incident to this point, they are taken under blindfold to an undisclosed area and held and debriefed. Other military personnel clean and secure the trail and eliminate all traces of an overnight camp. They also move their vehicles miles away and leave them at an archaeological site frequented by Holden near the town of Ruidoso.

How the Military or anybody else could have got into the area and the object out on to the main road, then back to wherever they were taking it without alerting everybody in the countryside and tearing up the whole environment is nearly as big as mystery as the whole incident. The only positive contributing co-factor to the endeavor is that the object could have been extremely light weight for its size. Even so, in spite of the questionable nature of what they faced, the site is cleaned and secured within 6 hours. The Military, not realizing there was another debris field on the Brazel ranch, in their efforts think they have contained any information of an unusual nature about the crash from leaking out.

In contrast to the archaeologist and firefighter stories, it should be mentioned that William J. "Bill" Edgar --who was a former farmhand and longtime resident of Capitan (he died at age 84 in December of 1998 in the city of Roswell) -- in a 1996 interview said in 1945 or 1946 he moved to Pine Lodge Road. He said he was there in July 1947 and continued to live there until 1991 when he moved to a rest home in Roswell. He said the incident never happened. He never heard about flying discs, saucers, or soldiers being there and until the time of the interview nobody ever asked him about it either.

It must be remembered however, that the object was reported down at 11:27 P.M. Friday night. The military was there by dawn Saturday and as stated in the paragraph just above "the site is cleaned and secured in 6 hours." Besides, not only being a 3-day Fourth of July weekend when lots of times lots of people go places, it was done so quickly and under a blanket of secrecy that Edgar could have easily just missed the whole thing.

Even though Edgar mentions Pine Lodge Road, he claims to be a long time resident of Capitan. Capitan is on the south side of the mountains while the crash occurred on the north slope. There is a 10,023 foot mountain between the 2 places. Now true, Edgar may have been using Capitan only as a mailing address because of the post office and may have been living in one of the small cabins scattered throughout the general area. However, there are any number of other post offices much closer to Pine Lodge than the one in Capitan. Also Edgar claims to have been a farmer. The crash site location is the Capitan Mountains, not known for having much in the way of farmland.

In a similar vein (again implying that no such crash occurred in the vicinity), some people have said it would be highly unlikely that such a spectacular crash would not have been seen or at least heard at Pine Lodge -- a then-popular log-cabin style nightspot located only a few miles away that has since burned down and alleged to have had a sizable crowd on that Fourth of July weekend. Especially so since the archaeologist and his students reported seeing the object come down from somewhere in the same general area.

It isn't exactly known where the archaeologist camped that night. But Pine Lodge had been built in a valley created by Boy Scout Mountain which is a peak that stands 6,726 feet high. It is in an almost direct east-west line between the crash site which occurred at about 2000 feet on the east and where the lodge to the west used to be. Which would make seeing or hearing the crash from the lodge highly improbable. Even if there was a huge or brilliant flash on the other side of the mountain from the lodge, it would have been sitting in the direct shadow from it, totally diminishing any effect caused by being in a close proximity.

In the fall of 1989, after seeing a televison program on the Roswell incident a woman contacted the producers identifying herself as Mary Ann Gardner, a former nurse at a St. Petersburg, Florida hospital. She told them that in 1975 a terminally ill cancer patient told her that at one time she was part of a group out looking for fossils. As the group was exploring the landscape, they came upon a crashed craft of an unknown type. No sooner had they arrived when an extremely large number of armed U.S. military personnel swooped down on them, apparently having stumbled into some sort of a secured area. The military swore everyone to secrecy and escorted them out of the area.

According to the nurse, the dying woman said she thought the location of the crash site was Mexico. Looking back, the nurse felt she meant New Mexico instead because the presence of such a large contingent of U.S. troops implied that the crash scene was most likely in the U.S. The nurse had no reason to question or clarify the location at the time because when she was being told the story it wasn't being connected to the Roswell incident. It was only after seeing the television program 14 years later that the information the dying woman imparted meant anything.

The nurse related that the woman told her that she was not an official part of the group but had gone with a friend. Whether they were professionals, amateurs, or students, she didn't say. Nor did she specifically mention the time frame the story unfolded. But Gardner felt it was most likely sometime in the late 1940s because the woman had related to her previously she was a student during that period, making a rather good case that her friend in the group was a student.

The nurse had access to the woman's charts when she was in the hospital and may have known her actual age at the time. But 14 years later, she just could not recall what it was. The woman looked to be in her 70s which would have made her in her in her 40s if the incident occurred in 1947. However, she was terminally ill, dying of cancer, and could have easily appeared to be 20 years older than she actually was which would make her anywhere from age 22 to 27 in 1947. Many people had put their lives on hold during the War years so it wasn't unusual for college students to be in their late 20s or even older at the time and certainly not unusual for graduate students. So age was not a factor except for being within a range to be feasible.

It should also be mentioned she said she was with a group hunting fossils. Holden and his students were looking for Native American artifacts and signs of pre-contact occupation, not fossils. However, the woman only joined the group to be with a friend so maybe at her level of participation, student, graduate student, or no, there wasn't any difference between an artifact and a fossil.

Why would the good Professor be anywhere near the Capitan Mountains over the Fourth of July holiday in 1947 in the first place? Good question. More than likely, Holden was following up on a lead or a tip. The 3-day weekend gave him a long enough break in his classroom academic routine to pursue it.

Prior to the outbreak of World War II, Holden was known predominately for working sites in Texas and the Arrowhead Ruin in New Mexico. Following the War, his attention shifted to southeastern and south-central New Mexico. He had been introduced to a previously unknown archaeological site (the Bonnell site) in the Hondo River Valley near the town of Ruidoso by Peter Hurd (a local artist). The Bonnell Site -- which Holden worked extensively after the War and fanned out from regularly in search of additional archaeological support data prior to his formal excavations he began directing in 1950 (some 3 years after Roswell) -- is only about 15-or-20 air miles from the Capitan Mountains crash site.

Holden operated the only college level student related operations in the general area. However, at the time (the summer session of 1947), Holden was teaching a history course at Texas Tech in Lubbock, Texas. On the long Fourth of July weekend, in that he was intimately familiar with the area, he could have easily been on a field trip to the Capitan Mountains. In-and-around the general area is known to be rich with an abundance of prehistoric archaeological remains including an accumulation of round, semi-subterranean houses with adjacent storage pits and trash deposits of food remains and other domestic debris stemming from the early Jornada Mogollon culture. Searching almost any part of the area for a minor or significant find even on a simple hunch rather than a lead or tip would easily fall well within the scope of reason. As a matter of course, a friend of one of the students traveling with him could have easily joined the group as well. See Roswell Archaeologists: The Dirt Before The Dig .

In an interesting side note, it is reported that Grady L. "Barney" Barnett (mentioned above saying he saw a 20-to-30 foot diameter disc that had crashed on the Plains of San Augustin) told his friends that he too had encountered an archaeologist with a group of students that came upon the scene at almost the same time he did. The students were said to be from either the University of Pennsylvania or the University of Michigan. (Although highly top-secret at the time, the Signal Corps Engineering Laboratories via University of Michigan AND Pennsylvania worked very closely with the Hermes B-1 rocket program. The July 3rd launch -- even though it was not designated as a Hermes vehicle -- was interestingly enough under the Signal Corps auspices).

The professor leading the group was said to be a Dr. Buskirk. Rumors had it the group was exploring cliff dwellings in the area. The 500-room ruin at Gallinas Springs is just 12 miles northwest of Magdalena, NM but wasn't widely known in those days. Although not cliff dwellings, the Bat Cave (which is an archaeological site of major significance) is located almost on top of the impact site.

In a somewhat incredible coincidence, it turns out there just happened to be a Winfred Buskirk working on his Ph.D dissertation at the exact same time in the desert southwest. Dr. Buskirk, an anthropologist who was a military officer in WWII and in the reserves for about 20 years, denied ever being in New Mexico in 1947 saying he was in Arizona all of July

1947. However, he was working in eastern Arizona near the New Mexico border only a short drive time from the Barnett site although he is on record as saying "I was certainly too busy on the reservation (Fort Apache Indian Reservation) to be engaged in any archaeological sideshow."

It must be said, over time there appears to have been a huge meshing together of these archaeologist sightings. The various people who write or report on all this either accept or reject the archaeologist theories. Those that accept seem to have taken bits and pieces of each story and combined them almost as one, which now undercuts the credibility of either or any of the possibilities being true.

The archaeologist at the Plains of San Augustin has NOT been pinpointed accurately even though he and the students were working, or at the very least visiting for the day, the Bat Cave, which is a site of major importance. Suggestions as to who the archaeologist might be runs the gamut from the wild and wooly amateur rockhound, Pothunter and desert rat William Lawrence Campbell (known as "Cactus Jack" and who claimed to have seen Foo Fighters during World War II), to the known or pinpointed archaeologist W. Curry Holden. It must be remembered, however, Holden never came forward until he was practically on his deathbed. And even then only after he was sought out.

Such revelations as being connected with the Roswell Crash: UFO Down have not always proven healthy on one's career. Holden's daughter Jane Holden Kelley is on record as saying that at the time the Roswell interviews were being conducted with her father, because of his age he was easily confused. Memories from his life were jumbled and reordered and even though she and her dad were close, he had never mentioned anything about the Roswell Incident.

Again, such revelations are not always healthy on one's career. Especially an academic. Even though the Bat Cave is a major site and being involved with it on a professional basis would be a feather in one's cap, not one archaeologist or anthropologist has stepped forward. A number have been considered including the aforementioned Larry Campbell)known as "Cactus Jack" brought to our attention by Thomas J. Carey) to such luminaries as George Allen Agogino, Art Jelinek, Jesse Jennings, Donald Lehmer, William Pearce, C. Bertrand Schultz, John Speth, Joe Ben Wheat, Ridgely Whitman, and Regge Wiseman.

Most have been rejected by UFOlogists for one reason or the other ranging from not being viable to being too young or inexperienced in 1947 to just not being able to be there for clearly provable reasons. The nearest to the location of the crash site that remotely meets the necessary criteria is Dr. William Pearce, a close colleague of Holden. He was working the Arrowhead Ruin that summer. The Ruin -- located between Glorietta and Pecos, New

Mexico -- is within easy shooting distance of the Plains. There is, however, no evidence that places Pearce anywhere near the crash site.

Barnett is different. What sounds like is going on with him is a combination of several things. As brought out about him previously, it is thought he was with students at the newly discovered Bat Cave where some of the oldest examples of maize (that is, corn) has since been found. The problem is Barnett has no reason to be with students as he was a government soil conservation engineer not an educator. More than likely in that he had intimate knowledge of the area and local terrain, he was either told to be a guide or recruited as a guide because the caves are quite remote and difficult to get to (and on BLM land).

The archaeologist and students were probably with him and they went to the crash scene together whether it was an errant V-2 or a disc or not. They were told not to say anything which would be the case if the crashed object was something as unspectacular as a regular V-2 that came down on public land outside the designated impact zone, a super-secret Hermes B-1, OR a disc. Eventually, however, the story got out about Barnett being there followed sometime later by the archaeologist and the students being there. Since the story got out separately they were put at the scene separately.

It seems like it is Barnett himself that has something to hide about the archaeologist. It may not be the deep dark or sinister secret it sounds like. As a matter of fact, it might be something as boring as Barnett simply moonlighting on the side and not wanting his wife to know he had an outside source of income.

Speaking of the archaeologist and students in Barnett's case, it could be the nurse Mary Ann Gardner (above) in talking about her patient and her experience coming across a crash scene with fellow students although it is unlikely (instead of being with Holden and his group at the Capitan Mountains site, she could have been talking about the Plains of San Augustin).

The most important part of the whole Barnett affair is not so much if there were archaeologists, if he ran across a disc, or if it was the errant V-2 but that the V-2 radar system is 'on'. Again, since the use of radar is an important early stage regarding the timeline of the Roswell Incident, throwing doubt on its use or undermining any credibility of its use would be of the utmost priority.

Debris Field at Mac Brazel Ranch

(directions)

Following the rain the night before, Mac Brazel on horseback in his usual routine inspects the pastures surrounding the ranch house. Riding with him is the young son of neighbors, William D. (sometimes Timothy) "Dee" Proctor, age 7. During the inspection, Brazel discovers a large debris field. Scattered on the slopes and into the sinkhole and depressions are metal, plastic-like beams, pieces of lightweight material, foil, and string. The debris is thick enough that the sheep refuse to cross the field and are driven around it to water more than a mile away.

Brazel takes a few scraps of the material and heads to the home of his closest neighbors, Floyd and Loretta Proctor (the parents of Dee Procter). He shows them "a little sliver" of material that he can neither cut nor burn. He tells his neighbors about other material on his property including what looked like aluminum foil saying it was very flexible and wouldn't crush or burn either. The material was later tagged "Memory Foil" because when crumpled it would straighten out, not stay creased, and somehow open out to its original shape as though it had a memory. The Proctors suggest he take some of it into town to show the sheriff.

Brazel goes into the little town of Corona to do some shopping and visits Wade's Bar and pool hall. There for the first time, he hears stories about the recent flying saucer sightings in Washington state and about an eyewitness account by a man that came through town that was run off the road by 2 flying discs in Arizona a few days before.

After returning to the ranch later that evening, Brazel removes the largest piece of debris from the ranch property he could find, reportedly about 4 feet long by 3 feet wide by 1 foot thick and "as light as a feather." Brazel loads it into his truck, storing it in a livestock shed (sometimes described as a "barn") some distance north of the crash site.

Sunday, July 6, 1947:

Brazel gets up early, completes his chores, and then drives into Roswell about 75 miles away. He stops at the office of Sheriff Wilcox showing him the scraps he gathered and telling him about the debris field. Wilcox -- putting together the phone call from the archaeologist and students the previous day regarding a crashed "aircraft" and now what Brazel is saying about a debris field on his ranch -- concludes the 2 events may be connected. In that it might be an airplane or possibly some sort of a classified aircraft belonging to the military because of the unusual nature of the debris, he contacts authorities at RAAF.

While waiting for the military officers to arrive, Wilcox dispatches two of his deputies to the ranch. They have only the directions given by Brazel. But both men are familiar with the territory and Wilcox believes they will be able to find the debris field. In the meantime, the police officers and firefighting crew including Dan Dwyer return from the crash scene at archaeologist site tired, dirty, and all upset over their treatment by the Military. They have not been debriefed at the level of Holden and his group because (although Dwyer secretly did) none of the firefighters including Dwyer or any of the police officers are suspected of having actually seen anything.

The 2 deputies dispatched earlier in the morning by the sheriff return back to the police station sometime mid- to late-afternoon. They report the ranch seemed deserted and even devoid of animals. They saw no sign of debris as they drove up to the ranch house from the main road which would be the case in that the debris field was some distance from the ranch house anyway. Nowhere is it recorded who they were. However, on the way back they noticed several army trucks and a number of armed soldiers that seemed to be staging at various places along the road.

Colonel William Blanchard, commanding officer of the 509th Bomb Group, either holding back information on the archaeologist site or not knowing about it because it was handled by White Sands orders Jesse Marcel, the Air Intelligence Officer to investigate. Marcel interviews Brazel, examines pieces of the material that Brazel brought in, and decides he had better visit the ranch and examine the field himself.

Marcel, taking some of the debris with him, returns to the base and reports to Blanchard what he has seen. Blanchard, convinced that he has in his possession something highly unusual and possibly connected somehow with the object found previously, alerts the next higher headquarters.

Marcel returns to the sheriff's office with the senior Counter-Intelligence agent assigned to the base (sometimes "the man in plain clothes") Captain Sheridan Cavitt. They escort Brazel back to his ranch to examine the debris field.

Brazel takes the 2 military officers out to the crash site. The debris field is three-quarters of a mile long and 200-to-300 feet wide. A gouge starting at the northern end of it extends to 400-or-500 feet toward the other end. It looks as if something touched down and skipped along. The largest piece of debris recovered (taken to the shed by Brazel previously) was found at the southern end of the gouge. Even though the gouge was fairly long and wide with some depth to it, it was not clear what part of the craft had done the damage. What was thought to be the main body of the craft or object was stumbled upon some miles away by Holden and his students. No sign of a power source was found as though the power source and the main body had become forcibly separated. And although the materials found at the Brazel site were extraordinarily light, the furrow must have been made by a very heavy component (possibly the power source before it exploded) spreading parts and debris all over field.

The remaining debris is as thin as newsprint but incredibly strong. There is foil that when crumpled unfolds itself without a sign of a wrinkle; I-beams that flex slightly and have some symbols on them; and material resembling Bakelite. Marcel makes sketches of the markings on the I-beams later describing them as Hieroglyphic Writing and thought by many to be alien symbols.

Marcel and Cavitt are there the rest of the day Sunday. They walk the perimeter of the field and then range out looking for more details of another crash site but find nothing else. Finally they return spending the night in sleeping bags in the "Hines" house (an old ranch house near the debris field) and having cold pork-and-beans and crackers for supper. The next day (Monday) is devoted to collecting debris.

Late in the afternoon they load the back seat and trunk of Marcel's 1942 Buick convertible and then the Dodge-jeep carryall driven by Cavitt. Shortly after nightfall as darkness closes in, they begin the trip back to Roswell. Sometime during the 2-day period, personnel from the local Roswell radio station KGFL arrive to wire record an interview with Brazel. The crew, realizing there is no electricity on the ranch (or being overseen to closely by Marcel and Cavitt), requests Brazel to return to town with them.

Dodge Jeep Carryall as driven by Cavitt. Most people think of a small-size jeep unable to carry very much. A Carryall is, however, a full-size truck-like vehicle.

William M. Woody, age 14, lived on a farm east of Roswell AAF. Several nights earlier, he and his father observe a brilliant fireball headed out of the northwestern sky that appears to come down northwest of Roswell. His father, convinced it is a meteorite, in the afternoon takes William in his flatbed truck to look for it.

About 19 miles north of Roswell on U.S. 285 where the highway crosses the Macho Draw, William sees a uniformed soldier stationed beside the road. Continuing, they come across military vehicles and more armed sentries -- some with rifles, some with sidearms stationed at ranch roads, crossroads, or any access leaving the highway including the Corona road (State 247) which runs west from Highway 285. Traveling as far as Ramon 9 miles north of the 247 intersection are sentries, too. The report is that the road is similarly blocked all the way north to Vaughn and highway 60, the same road that runs through Fort Sumner. His father turns the truck around at Ramon and they head south toward home.

Notice the timing on this. Brazel gets up early, completes his chores, and then drives into Roswell about 75 miles away but by the afternoon of the same day when William M. Woody and his dad head north from Roswell all the side roads are guarded by armed military personnel. When Brazel left early in the morning, neither was he stopped nor did he report any sort of encounter with anyone from the military. Remember as well, the 2 deputies sent to the debris field early in the morning by Sheriff Wilcox returning to the station-house around mid-day. On the way back, they reported seeing several Army trucks and a number of armed soldiers that seemed to be staging at various places along the road. KGFL people picking up Brazel made no mention of a military presence, however.

A scientist friend of W. Curry Holden's -- not an anthropologist but a vertebra paleontologist by the name of C. Bertrand Schultz -- reports he saw the road blocks and military presence as he drove north from Roswell over the Fourth of July weekend. In December 1947, Schultz presents at the 46th Annual Meeting of the American Anthropological Association held December 28-31 in Albuquerque. Holden attends the conference. Coincidently, the bio-searcher (discussed below) hoping to hear Ruth F. Kirk present Aspects of Peyotism Among the Navajo just happens to attend the same conference as well. Knowing Holden, the three meet up.

Some reports have Schultz going from Nebraska to Roswell in July, 1947 rather than the other way. But in conversation with Holden at the conference, he brings up the fact that he had tried to meet with him in the field over the Fourth of July holiday earlier that year. Schultz had been told that Holden was going to be at the Bonnell Site over the long weekend. Since he was on his way to Nebraska and had the time (although it was out of the way), he thought he would go through Ruidoso to see what Holden was up to. There he was told Holden had taken a group of students on a field study near Roswell.

Unable to find Holden, Schultz continues on to a site called the Arrowhead Ruin -- an Indian pueblo dating from circa 1370 to 1450 located south of Santa Fe, New Mexico between Glorietta and Pecos, then being worked by another colleague William Pearce. He sees the military cordon on the western side of the highway as he heads north out of Roswell. Since his eventual destination is Nebraska (although he grows more curious as he continues northward), he isn't excessively over concerned about roads being blocked toward the west one way or the other. At the most, at the time not knowing of the Roswell Incident, he attributes the military presence to no more than an exercise of some type. Only in retrospect did any of it take on any meaning for Schultz.

Did Holden reveal the nature of what he saw to Schultz during their meeting at the conference? Conjecture would indicate the answer is yes in that it was Schultz that brought Holden to the public's attention initially. (see)

Lt. Col. Philip J. Corso -- eventually to become an intelligence officer on General Douglas MacArthur's staff during the Korean War and a close friend of SAC commander General Curtis Le May -- was stationed at Fort Riley, Kansas in the summer of 1947. In 1997, a book he writes is published titled The Day After Roswell. In that book, Corso states that on the afternoon of Sunday, July 6, 1947 (2 days after the purported crash of the mysterious craft outside Roswell), five two-and-a-half-ton trucks and side-by-side low-boy trailers pull into the base loaded with huge wooden crates. Freight records indicated the crates contained aircraft parts (supposedly landing gear struts for P51s) coming from Biggs Army Air Field, Fort Bliss, Texas and bound for the Air Material Command at Ohio's Wright Field. Unusual in that aircraft parts typically flowed the other way (that is, from rather than to Wright Field).

Corso goes on to say that same evening he was serving as post duty officer and an enlisted man he knew through a local bowling league was posted as a sentry that night. When Corso approached the enlisted man's post, the sentry tells him the crews of the deuce-and-a-halfs have their own security and they told him "they brought these boxes up from Fort Bliss from some accident out in New Mexico".

Corso also writes from his role as post duty officer he located a routing slip related to at least one of the crates which indicated it was from a craft that had crash-landed near Roswell earlier that week. It should be stated, however, if the convoy was under a cloak of top security then any sort of an accurate (what Corso calls a) routing slip would most likely not be available or even exist. If it did, at the very least it would be a phony or counterfeit cover document of some sort such as the mention of "landing gear struts for P51s".

Upon arrival the convoy was most likely secured in a remote section of the base and -- as post duty officer -- Corso had an inkling of something going on. Meals to the crews had to be provided, showers, etc. so he probably received his information BS'ing with convoy security, the truck crews or possibly even cooks or food servers, then reported in his book he saw a routing slip so it would sound more official. No mention is made of actually observing a classified or top-secret shipping or travel manifest.

On that exact same Fourth of July weekend, the young boy mentioned previously in the opening paragraph is traveling with his Uncle. His uncle -- a bio-searcher who because of his discoveries will eventually have several plant species named after him -- is the same man who 13 years later in 1960 becomes notorious as the informant in the series of books about the Yaqui shaman-sorcerer Don Juan Matus written by Carlos Castaneda.

The boy and his uncle just happened to be on a road trip so the boy can learn firsthand about The Long Walk endured by the Navajos and Apaches as well as visit the gravesite of Billy the Kid located 2 miles outside of Fort Sumner, New Mexico after an excursion into the Arizona Strip. The two had been searching for fossils related to the Teratorn, a giant bird with over a 20-foot wingspan thought to be the inspiration of Native American Thunderbird legends.

Leaving the Arizona Strip in an impromptu decision following an indepth discussion about the Long Walk, they decide to go to Fort Sumner and while there visit Billy's gravesite. During the trip the boy gets a touch of food poisoning and spends a long uncomfortable night awake. Monday morning June 30 finds them on old Route 66 just south of the Grand Canyon outside Williams, Arizona. They are headed in the direction of the Elden Pueblo where prehistoric Native Americans had buried an extreme rare type meteorite in a ritual fashion and then on to Meteor Crater.

Sometime between 9:00 and 10:00 AM, the boy dozing on and off because of lack of sleep is startled awake by his uncle swerving the truck and yelling something like "What the..." The boy is thrown to the floor of the cab and because of same is not able to see the cause for concern. According to his uncle, swooping in behind and over his left shoulder from out of the northwest and only a few hundred feet above the top of the truck were 2 large sharp-edged almost flat circular-shaped objects, blunt across the back and seemingly made of metal. The objects were flying side-by-side with one slightly in front, both headed ESE out over the horizon at an incredible high rate of speed.

In only the few seconds it takes the boy to scramble up from under the dash, the objects are gone, leaving in their wake only a small residue lingering in the air like the smell of electricity and a quarter-mile wide swath of thick swirling air laying turbulently above the treetops like a sweltering mirage over a desert dry lake.

Interestingly enough, just a few short days earlier, on Tuesday, June 24, 1947 23 miles WSW of Mt. Rainier, Washington, the FIRST widely reported breakthrough UFO sighting in the U.S. press occurred. The infamous sighting by Kenneth Arnold of 9 disc- or saucer-like objects flying in a chain formation and estimated to be traveling at the then unheard of speed of over 1200 mph. When the bio-searcher fly-over occurred outside the little town of Williams, Arizona, the boy and his uncle had been in the back-country for close to 2 weeks and had not yet heard of the Mt. Rainier incident. Yet the verbal description of the objects seen by the uncle and Arnold are astonishingly similar in both cases.

On one of the days of the 3-day Fourth of July holiday, because the Long Walk study and the visit to Billy's gravesite was not part of their original itinerary and there is a long travel day ahead, the boy and his uncle are up and on the road early. Without any knowledge of the Roswell Incident at the time, after breaking camp close to daybreak as the boy and his uncle are about to turn left onto a main highway from some side road not far from Fort Sumner, they are stopped by a military convoy. Fort Sumner -- located on Route 60 about 55 miles east of Vaughn, New Mexico although not connected directly by main roads to the little town of Ramon -- is less than 40 miles northeast of there. Ramon is the same place that William M. Woody and his father (mentioned above) decided to turn around and go back from because of all of the side roads being blocked by armed military personnel all the way from Roswell.

The convoy -- itself headed north or northeast -- is composed of several flatbed trucks carrying large crates, some covered with tarps some not, escorted by jeeps and followed in the rear by a huge tow truck. The uncle makes his left turn and eventually catches and passes the convoy, continuing on their trip without incident.

However, the event is highly memorable for the nearly 10-year old boy. He had witnessed the Howard Hughes flying boat being moved in a similar fashion and just the sight of all the Army trucks trundling along out in the middle of the desert was exciting. But passing them, smelling the diesel, hearing all the noise, seeing all the wheels, and having the drivers salute or give a wave going by is unforgettable. The uncle does mention to a friend over dinner some days later (which meant nothing to the boy at the time) that he thought it was highly unusual that all the unit designations on the bumpers had been blanked out or painted over (see).

The convoy witnessed by the boy and his uncle outside Fort Sumner by the way and headed toward Kansas is almost an exact duplicate of the convoy as described by Lt. Col. Philip J. Corso in his 1997 book as arriving at Fort Riley, Kansas sometime in the afternoon of Sunday, July 6, 1947. In regards to the convoy, even though William M. Woody and his dad had been on the road between Roswell and Ramon on the same afternoon, nowhere has it been reported that they saw any sort of a military convoy. The implication being that for the convoy to have arrived in Kansas in the afternoon, it had to leave New Mexico during the night or fairly early on that day.

Brazel was on the exact same road during the early morning hours the Williams' were and he never reported seeing any sort of a convoy. Whatever is being transported in all those crates and under the tarps sighted outside Fort Sumner most likely came from the archaeologist site in the Capitan Mountains rather than anything from the Brazel debris field. If you recall, a full day before mid-morning July 5, 1947 at the archaeologist site, Roswell city firefighter Dan Dwyer saw what he describes later as a "strange craft" being lifted into the air by a crane and set on a flatbed truck, then being secured with chains and cables and covered by a tarp. It appears the convoy must have headed northeast on Highway 54 out of the Capitan Mountains to Vaughn then cut across New Mexico east toward Kansas on 60 through Fort Sumner.

Why some crates are shipped across country via flatbed truck to Wright Field while other crates are selected to be flown not only into Wright but also Fort Bliss and Fort Worth is not known. At the time -- at least for day-to-day "official" traffic -- Wright Field was reportedly closed down for modernization. However, it was still operational on a contingency basis. The Army does rely on a built in redundancy.

Perhaps as a matter of course, they trucked some material and flew some material. It could be a diversionary tactic, weight, the nature of expediency of whatever was in the crates, the non-availabilty of aircraft, not enough cleared personnel, or even inter-service inter-agency rivalry. It could be smaller crates were just short-hopped to Fort Bliss and Fort Worth and then for some reason off-loaded and collected or possibly assembled and placed into larger crates of similar or related material onto the side-by-side low-boys for Wright.

Between 1942-1945 a massive construction effort was begun at Biggs Field at Fort Bliss. Huge hangars and some of the longest concrete runways in the world were built to accommodate heavy bombers. It could be one of the huge hangers was used to assemble parts or wreckage into a more complete craft similar to what the FAA does with a crashed airplane. After all, at least one of the crates was reported to be 20 feet long by 4 or 5 feet wide and 4 or 5 feet high.

Where all the boxes and crates went to -- that is, where they were eventually stored or ended up or the parts and materials used for despite all the speculation -- is not known (see).

Before moving on, it should be pointed out the flight path of the 2 disc-like objects that flew directly over the top of the bio-searcher's truck outside Williams, Arizona coming out of the northwest and headed ESE -- if they continued in their slightly curved trajectory -- their path would have taken them right over the Fort Apache Indian Reservation that Dr. Winfred Buskirk claimed to be working.

Continuing on the same eastward curve slightly toward the north, at the ultra high speed they were traveling the objects would be over the Plains of San Augustin within a matter of seconds. Coincidently, that very same Monday morning June 30th disc sighting by the bio-searcher was the day just before the start of the July 1, 1947 RAAF and White Sands radar tracking cited at the top of this page stating: "Radar operators...while on routine duty begin tracking an object that defies convention."

ROSWELL DAILY RECORD, TUESDAY JULY 8, 1947

Tuesday, July 8, 1947

At 12:00 A.M:

Marcel stops at his house while Cavitt apparently continues on his way to the base in his own vehicle. Marcel awakens his wife Viaud and son Jesse Jr. to show them the material. Over the next hour they examine the debris on the kitchen floor. Marcel Sr. says it was a flying saucer.

With the help of his son, Marcel loads it into the car to be taken to the base. At the time, Marcel seems to think he is not breaking regulations since nothing to his knowledge has been classified yet. But even though Marcel is the Air Intelligence Officer at the base, he seems patently unaware and uninformed as to the level of or connection to the 2 events as the other event was highly classified (to wit, all the security and armed soldiers for example) almost as though up to that time the 2 events had not been connected on an official level OR Marcel had been left out on a "need to know" basis.

Higher-ups, however, have put the 2 incidents together and things change quickly as pieces from the Brazel debris field are combined and shipped with much larger pieces brought in from the other site and loaded onto waiting aircraft. The first flight is flown to Fort Worth Army Air Field by Colonel Payne Jennings, deputy base and group commander at Roswell Army Air Field. The second flight is flown by Captian Oliver W. "Pappy" Henderson, First Air Transport Unit, and co-pilot of the first B-29 flight from Roswell to Fort Worth. He also was the pilot of a C-54 flight carrying wreckage from Roswell to Wright Field, Dayton, Ohio even though it has been said Wright Field was closed to official traffic.

Sergeant Robert Earl Smith, also of the First Air Transport Unit, is a member of the team that loads crates onto the various transport planes. Most of the crates are 2 or 3 feet high by 2 feet square. Years later, Smith reports that one was much larger -- about 20 feet long by 4 or 5 feet wide and 4 or 5 feet high. He also states the loading operation was supervised by a number of people in civilian clothes.

The loading crew was told that the crates were from a plane crash although as far as Smith could recall crashed planes were usually not flown out in crates under armed guard. It is also reported (albeit not by Smith) at least some of the crates remain on low-boy flatbeds and not off-loaded. However, the reason a definitive answer is not known is because in order to ensure that nobody except those who had a" need to know" gained access to the larger picture, the flight crews and ground crews of the various flights are kept separated with different loading and departure times spaced apart with different loading-area, unloading areas and landing locations. By now though, Marcel "gets it" and is flown to Fort Worth Army Air Field with a few wrapped packages of debris self-carried on the plane by himself or his staff personally.

According to Smith, during the loading operation (although it is not known how) one of the loading teams or a member thereof somehow comes across a foil-like fragment between 2 and 3 inches square. The unusual material has jagged edges and could not be permanently folded or creased. Even after crumpling it up, it unfolded itself back to its former shape. Since then, Smith has been reported as saying that despite the armed guards one of his colleagues was able to stash a piece of the strange fragment into his pocket.

None of the crates are described as being large enough to hold a fully intact disc 20 or 30 feet in diameter (which is interesting if the object was made out of material that couldn't be cut). Width, height, and diameter of any object too large would limit the ability of it being placed into the hold or cargo bays of any of the airplanes cited as being used in the transportation process.

The same would be true regarding highway and truck routes because of the width of bridges and heights of underpasses. Neither Corso or the bio-searcher ever said anything regarding the convoys they saw transporting anything that appeared to be unusually high, wide, or excessively heavy. Which might indicate the object either came apart or possibly delta or wedged shaped as some have reported rather than being disc or circular shaped.

The bio-searcher said that in the course of his assistance in the investigation of the object's trajectory cited further on that he heard rumors of the object being stripped of every possible thing that could be taken off of it or out of it. Thus stripped the bulk of the skeletal remains, framework, or superstructure too large or heavy to be transported without drawing undo suspicion was dragged across the highway and out into the desert flatlands just north of the impact site, put into a temporary shelter, then simply buried out of sight of prying eyes for later retrieval like some broken down abandoned truck left to rust out in the middle of the desert off some side road. If retrieval of same ever occurred, it is not known. It could still be buried somewhere out in the desert north of the Capitan Mountains for some unknowing prospector, rock hunter, or archaeologist to stumble across.

Getting back to Jesse Marcel (the Air Intelligence Officer at RAAF), regardless of what one may think regarding him and where he was in position to what was classified or secret information and what wasn't (or if he should have taken debris home or shouldn't have) remember on July 8, 1947, a press release stating that the wreckage of a crashed flying disk had been recovered was issued by Lt. Walter G. Haut, Public Information Officer at RAAF under order from the Commander of the 509th Bomb Group at Roswell, Colonel William Blanchard. It was printed in the Roswell Daily Record in the now infamous front page as shown above.

More than likely, initially there are 2 separate military agencies or units at work here, each in the beginning not aware of the others involvement. The archaeologist site was probably handled by personnel from White Sands. It was most likely through their radar system (run by Signal Corps Engineering Laboratories of the Office of the Chief Signal Officer and not the Army Air Force) that the downing of the object was first detected. It is most likely as well that it was civilian and military personnel from White Sands (possibly even a V-2 quick-strike recovery team) that first showed up at the Pine Lodge Road impact site.

The Brazel site on the other hand was stumbled on by a civilian in the process of his early morning chores and in turn reported to the local sheriff. The sheriff then handed it off to authorities at the Air Field. It is pretty clear that mixed signals were being sent out by an unknowing high command so it is within reason that Marcel, even though he is base AIO, might not have known the full status of the situation.

When Marcel took the debris home, it was just past midnight in the early morning hours of July 8th, the same day the Roswell newspaper article came out. So at the time, how could any of the information or material be considered any sort of a "military secret" in either the classical or implied sense. It is sure no secret that the article is published. It appears that someone changed their minds while Marcel was in the field or in transit.

If you recall, both he and Cavitt were at the Brazel ranch for nearly the better part of 2 full days with no electricity or telephone communication or at least anyway until the arrival of a further military presence. You can tell by the whole tone in deliverance between the newspaper of the 8th and that of the 9th there has been a huge change!

After returning to base from the Brazel debris field, either Marcel or Cavitt (and it was most likely Cavitt) reports that Brazel left the ranch with KGFL personnel to participate an interview. Brazel is taken into custody almost immediately and kept under wraps by the Military for a week. Brazel is according to Major Edwin Easley, the 509th Provost Marshal, "assisting" the Army Air Force while staying in "the guest house" on the base. From the outcome seen the next day, most likely military officers are in the process of convincing him not to say anything about what he has seen and trying to keep him out of the way of reporters.

The TWO MOST IMPORTANT factors in the whole Roswell affair and series of events that most people miss are:

• the Brazel debris field got out of the bag to the general public

• the archaeologist site had not

At the time of the event and sometime afterwards the archaeologist site remained for the most part pretty well contained. To cover their tracks, a whole series of intentional disinformation was put into place to ensure any questions about a flying disc was directed toward the Brazel site, blurring both locations into being only one -- Brazel's. Why?

Because there was no sign of an actual craft at his place. Only scattered debris that could easily be foisted off as a weather balloon. Heavier or larger components such as a power source (if there was one) have already been crated up and carted off or remain laying out on the desert floor some distance away from the debris field along the object's flight path, unknown and undetected while the archaeologist site is in the final stages of being totally sanitized.

There is some question as to if Colonel Blanchard was still in charge or actually even the base commander at Roswell at the specific time the now infamous July 8, 1947, press release (shown above) was authorized for distribution. Special Order Number 9 issued by Headquarters, 509th Bomb Group and dated July 8, 1947 referring to a TWX (teletypewriter exchange) dated Sunday, July 6, 1947 reads as follows:

"Pursuant to the authority contained in Hqs. 8th Air Force TWX number A1 1593 6 July 1947, the undersigned hereby assumes control of the Roswell Army Air Field, Roswell New Mexico. Effective this date." The Special Order was signed by Payne Jennings, Lt. Col. A.C. (Air Corps), commanding.

The special order would seem to indicate Blanchard was not the commander in charge. However, the morning report indicates that Blanchard was present for duty July 8th but signed out on leave prior to the July 9th morning report. If Blanchard signed out the afternoon of July 8th, then the morning report would show him present on the 8th and gone on the 9th. Which is what the report shows. Other than who was responsible for the press release, Blanchard or Jennings or even someone with a bone to pick, the fact remains it was released.

The military is actually pulling a fast one. Top brass know it would be all right for Blanchard to go ahead and continue on his leave as scheduled IF all the ruckus was over something as simple or as trivial as a downed weather balloon. If it was something much more significant -- say a flying saucer for example -- then of course he couldn't. So they go ahead and send him on leave. That way it gives the appearance that the event is not important (at least not important enough to have him stay around).

In the scheme of things, the July 6 TWX becomes as important as the Special Order. The Military is responding to the events of the day before because they know WHAT has been found at BOTH impact sites on July 5th, especially so what has been found at the tightly kept under wraps archaeologist site.

The fall guy is Lt. Walter G. Haut. The big guys get away. Years later in a signed affidavit, Haut states that in July 1947 he was stationed at the Roswell Army Air Field serving as the Public Information Officer. He states that at approximately 9:30 AM on July 8, he received a call from Blanchard who Haut refers to as the base commander in the affidavit. Blanchard tells him he has in his possession a flying saucer or parts of one. He said it came from a ranch northwest of Roswell and that the base Intelligence Officer Major Jesse Marcel was going to fly at least some of the material to Fort Worth.

In a side note, photos of Marcel holding material identified as part of the debris he found shows foil that appears to be wrinkled, torn, and creased. Not exactly what one might expect from "memory foil". Marcel in an interview with Linda Corley in 1981 (his last) stated that Eighth Air Force Commander General Roger Ramey had him cover up the real material before the photo shoot. However, recent digital scans of photos taken in General Ramey's office on July 8, 1947 by James Bond Johnson (a reporter-photographer for the Fort Worth Star-Telegram) show that the debris displayed is consistent with eyewitness reports and is not a weather balloon which had been substituted on orders of the General. This includes clear identification of the hieroglyphic symbols displayed on I-beams and very thin super strong foil-like material that resisted crumpling.

Wednesday, July 9, 1947:

 Morning newspapers say the story that the "flying saucer" found near Roswell is a weather device. Some quote Ramey while others quote "informed" sources including senators in Washington. Reports are military officials tour news media offices in Roswell, Albuquerque, and Santa Fe retrieving original copies of the press release sent out by RAAF that revealed the Army had a "flying disk" in their possession. Mac Brazel remains in "custody" under the auspices of military authorities. A wire recording interview he made for the local radio station KGFL is conveniently misplaced or somehow mysteriously disappears. He is not released by authorities until July 15th (see).

Cleanup of the archaeologist site near Pine Lodge Road is already completed. Cleanup on the Brazel site resumes at sunup. The Military is trying to get everything picked up before any more civilians stumble across the field. Before they get the chance, Bud Payne -- a rancher in the Corona area chasing a stray cow -- crosses onto the ranch. No sooner is he on the property than a jeep with several armed soldiers pull up and escort him off the property.

Thursday, July 10, 1947:

Bill Brazel learns about his father's activities in Roswell as he is reading the morning newspaper. He realizes that no one will be at the ranch and makes plans to get down there to help.

At the debris field and impact site, men are working to get everything cleaned up. They want nothing left and no signs of their presence. The onetime 400-or-500 foot gouge at the Brazel ranch is now gone, smoothed over and restored to near original pre-crash condition. They reintroduce and release livestock into the area to wander around and track it up as much as possible. To the casual observer, it looks as though nothing had ever happened.

Dr. Lincoln La Paz -- who had thousands and thousands of hours of scientific time observing celestial objects himself - reported on July 10th, 1947 (6 days after the Roswell crash) seeing a huge elliptical-shaped object flying in the sky near Fort Sumner, New Mexico while driving by car with his wife and children. He saw a luminous unknown object sort of oscillating beneath the clouds. Its brightness was stronger than the planet Jupiter and its shape regular and elliptical. The nature of this object was unknown to the astronomer.

In a Life Magazine article dated April 7, 1952, La Paz is quoted as saying the object "... exhibited a sort of wobbling motion" and then disappeared behind some clouds. It reappeared and "projected against the dark clouds gave the strongest impression of self-luminosity."

The object then moved slowly from south-to-north and two-and-a-half minutes behind a cloudbank. According to La Paz's calculations (confirmed by his wife), the object was huge. As large or larger than the infamous "Battle of Los Angeles" object as presented in UFO Over L. A. seen by thousands in February, 1942 being some 235 feet long and 100 feet thick. (NOTE: according to reports as cited in the above link, the Los Angeles UFO was, however, thought to be closer to 800 feet in length). La Paz reported the horizontal speed of the object he observed ranged between 120 and 180 miles per hour and its vertical rise between 600 and 900 miles per hour.

(please click)

LITTLETON vs THE WANDERLING:

Battle of Los Angeles or UFO Over L.A.?

Friday, July 11, 1947:

The debriefings of all the participants are under way. Participants are taken into a room in small groups and told that the recovery is a highly classified event. No one is to talk about it to anyone. Everyone is told to forget that it ever happened.

Saturday, July 12, 1947:

Bill Brazel arrives at the ranch but no one is around. Brazel begins his work, first surveying the ranch to see what needs to be done. He sees no evidence of a continued military presence. The trucks, jeeps, soldiers, are gone.

Tuesday, July 15, 1947:

Mac Brazel is released from "custody" and returns home. All he will say about his experience is that his interrogators kept asking him the same questions over-and-over again and that Bill is better off not knowing what happened. Besides, Mac has taken an oath that he will never reveal in detail what he saw. By now, most of the World has forgotten that a flying saucer supposedly crashed in New Mexico.

AUGUST 1947

One Month after the Crash

A young ranch-hand hired by Brazel one month after the crash named Tommy Tyree is told by Brazel that even in the 90 degree July heat. his fully wool-covered sheep were too frightened to go to water on their own and he was forced to circle them a mile-or-more around the debris field to water because they refused to cross the area.

That same month Brazel and Tyree on horseback doing routine ranch maintenance spot a piece of metal debris from the wreckage in the water at the bottom of a sinkhole. Neither man bothers to retrieve it. The whereabouts of that particular piece of debris or in whose hands it eventually fell into has never been made public.

SEPTEMBER 1947

2 Months after the Crash

Lewis "Bill" Rickett is assigned to assist Dr. Lincoln La Paz from the University of New Mexico. La Paz's assignment is to determine, if possible, the speed and trajectory of the craft when it hit.

The problem for La Paz is twofold. First, he is used to figuring out the trajectories of meteors so his approach is somewhat unorthodox compared to determining flight paths of aircraft. Secondly, the recovery teams have done their jobs so well eliminating or smoothing over the sites and creating reluctant witnesses it is difficult to pinpoint an accurate trajectory initially.

Although the gouge on the ranch was reported to have been visible for at least a few years after 1947 by several eye-witnesses including Mac Brazel's son Bill and then Lt.-Colonel and now retired Gen. Arthur Exon who flew over the site in 1949, it is not readily apparent at the time from the ground. Because of such, over the heated objections of Rickett, La Paz -- who has a top-secret clearance from his World War II job at the Proving Grounds -- brings in a mysterious bio-searcher who knows Southwest indigenous plants intimately. Although the bio-searcher does not have anything close to a security clearance, he is a longtime trusted friend of La Paz and known to have an even longer working relationship with Albert Einstein. Since La Paz has carte blanche over the operation, there is not much Rickett can do about it except harbor hard feelings.

The idea is to have the bio-searcher determine if and where any of the growth may have been moved, removed, or replanted (see). It is a long and time consuming job. But the length and width along with the direction of the gouge is roughly figured out. Mine detectors are brought in to see if any metal debris or parts have been entered below the surface or in the surrounding area.

Retracing several miles in both directions of the suspected trajectory, both in the air and on the ground in an effort to confirm their conclusions, they discover a previously unknown and unspoiled touchdown point 5 miles from the debris field where the sand has somehow been crystallized. The plants and scrub brush growing along the periphery of the glass-like sand and gravel are not so much burnt or scorched as they are more-or-less trying to return to a natural growth stage after being severely wilted, apparently from whatever crystallized the sand 2 months earlier.

As well, the top portion of the sand and gravel in a definite north-south orientation in the major width between the scrub brush seems to give off a very slight practically non-observable blue hue in the bright sunlight. The hue is caused by what appears to be a transparent turquoise-like patina almost as though a fine veneer or micro-thin spray had fallen over the top surface of the sand (see).

Rickett and La Paz speculate the craft landed "cold" (without power) and then took off under power, exploding seconds later over the Brazel ranch. The bio-searcher after visiting the archaeologist site and finding a fairly well executed attempt at returning it to its natural state is convinced -- in spite of that attempt to camouflage the damage, something with some weight to it or at least speed, and apparently large enough to break limbs in a fairly wide track as well as being hot enough to scorch the trees and foliage -- angled through a top portion of the forest and down into the open area surrounding the boulders, ending up against the rocks.

But he is not convinced it necessarily means the object was extraterrestrial. No physical evidence such as metal pieces or scraps attributable to the object was found anywhere near or around the site. Nor was there any sign of the pale blue patina on the soil as observed previously at the fused-glass site. As to any sort of a military presence ever having been in the area, the only thing that would remotely indicate such a possibility was the finding what is called a P-38 can opener (a one and a half inch item issued to GIs to open ration cans).

Working with the Rickett and La Paz hypothesis regarding the fused sand site, the bio-searcher in observing the plant lean-bias and reading the surface direction of the heat bite on the leaves and stalks and discounting the possibility of fulgurites caused by lightning strikes in the sand suspects an extremely hot but very quick touchdown or possible low pass-by by the object, suggesting it may have been (except for some minor lift ability) out of control or not even controlled.

He is almost certain the debris pieces at the Brazel ranch belong to a totally separate object. But even so there still remains the possibility that this one was shedding or jettisoning similar parts or even breaking apart as it went along. Having either lost its power source or not having one (thus not being able to change speed, direction, or climb sufficiently), the main body possibly still traveling hundreds of miles per hour crosses the basically flat desert terrain within seconds, all the while radiating heat. Failing in an effort to gain sufficient altitude to clear the crest of the oncoming mountains, the object bounces hard into the short rough upslope of the landscape with a forced reduction of speed through the trees and dirt, sliding sideways into the rocks and boulders on the side of the Capitan Mountains some 35-or-so odd miles to the south southeast.

The Capitan Mountain site seems untouched on first glance with only hard-to-find minor scars and difficult signs of reworking found only after much scrutiny. La Paz knows 2 months earlier that the debris field had a gouge three-quarters of a mile long and 200-to-300 feet wide starting at the northern end and extended to 400-or-500 feet toward the other end. When La Paz and his investigation team arrive at the debris field, it appears normal at eye level.

Such is not the case a year-or-so after weathering. Because of short-term ground collapse, the gouge on the ranch becomes visible at least a few years after 1947 as seen by several eyewitnesses. They included Mac Brazel's son Bill and then Lt. Colonel and now retired Gen. Arthur Exon who flies over the site in 1949. It is reported in some circles that after Exon's flight when it became apparent that the inititial reworking of the debris field by mostly untrained and amateur military personnel (read, GIs) was not holding at the level anticipated, sometime post-1949 a secret, small, and very specialized scientific landscape geologic rebuilding team is brought in to counterfeit and encompass a much wider area in order to withstand almost any later post- pre-crash type scrutiny or at least render any outcome inconclusive (see).

The Cloud Shaman

At first Rickett, although not questioning his own hypothesis, is however not convinced that the fused-glass site and the debris field on the Brazel ranch are related even though all 3 sites (if you include the archaeologist site) are in an exact direct line with each other.

However, a young boy traveling with the bio-searcher (for unknown reasons) while busying himself looking for horn toads and lizards in the surrounding scrub brush and sandy terrain as well as breaking up rocks for the first time with a newly aquired prospector's pick comes across a few pieces of the same foil-like material as found previously at the debris field.

Rickett, who never particularly liked the bio-searcher (primarily because it always seemed he was right), quickly gathers up the pieces and in a rather harsh and abrupt fashion orders the boy and his uncle back to the vehicle they arrived in, placing them under guard with orders not to let them leave. In a widening search, Rickett finds more of the foil-like material and possibly other scraps and debris but doesn't let anybody see it or handle it. When Rickett returns to the truck, he finds the bio-searcher and the boy gone and the guard assigned to watch them having no clue where they went or what happened to them. A search of the area shows no sign of either the boy or the bio-searcher in the vicinity as though they simply disappeared or vanished, the desert and the surrounding environment somehow swallowing them up without a trace (see).

In the days that follow, La Paz interviews a fairly large number of people (many of them Spanish speaking) in-and-around the general Roswell-Corona area (especially along the suspected flight path trying to paste together facts). But he finds most people interviewed appear somehow intimidated in responding.

Memory Foil

Years later in a meeting unrelated to any of the above, a discussion between La Paz and the bio-searcher turns to the events at Roswell. The bio-searcher retrieves a small wooden box much smaller than a cigar box and hands it to La Paz. La Paz opens it and inside is a piece of the foil-like material folded to about the size of a matchbook.

The bio-searcher informs La Paz that unknown to Rickett or anybody else that day, the young boy had found and stuck the piece of material in his pocket, the piece eventually ending up in the hands of the bio-searcher. La Paz -- who had seen similar material under very strictly controlled circumstances -- is visibly surprised that the bio-searcher has such material. He sets the box down and redirects the conversation to another subject.

With the lid open, the foil mysteriously unfolds itself out of the container to its full extent (about the size of a small handkerchief). Completely covering the box, the paper-thin foil displaying no sign of folds, creases, or wrinkles. With no visible interest one way or the other by La Paz except the initial surprise, the bio-searcher returns the foil to the box with no further discussion.

What happened to that piece of foil or where it is now is not known. Some speculation has it that following the bio-searcher's demise because of a perceived relationship to Sarira, that it was possibly co-joined into the fate of the feather mentioned in Meditation Along Meteor Crater Rim. Others suggest that once it became known that a piece of the material was in private hands, it was appropriated by authorities.

Col. Philip J. Corso (cited above in regards to seeing the convoy arriving at Fort Riley, Kansas in the summer of 1947) writes the following in his book The Day After Roswell of the same or similar type foil:

"There was a dull, greyish-silvery foil-like swatch of cloth among these artifacts that you could not fold, bend, tear, or wad up but that bounded right back into its original shape without any creases. It was a metallic fiber with physical characteristics that would later be called "supertenacity". But when I tried to cut it with scissors, the arms just slid right off without even making a nick in the fibres.

"If you tried to stretch it, it bounced back. But I noticed that all the threads seemed to be going in one direction. When I tried to stretch it width-wise instead of length-wise, it looked like the fibers had re-orientated themselves to the direction I was pulling in. This couldn't be cloth. But it obviously wasn't metal. It was a combination to my unscientific eye of a cloth woven with metal strands that had the drape and malleability of a fabric and the strength and resistance of a metal. I was on top of some of the most secret weapons projects at the Pentagon and we had nothing like this even under the 'wish-list' category."

Since Corso wrote the above ("wish-list" or otherwise) in Discover Magazine, Vol. 25 No. 04, April 2004 a full 57 years after the mysterious crash at Roswell, Brad Lemley writes of a present day actual in reality known metal strip 8 inches long, 1 inch wide, and as thin as aluminum foil he saw that acts very similar to the late 1940s era memory-foil.

In the article, Lemley writes that William Johnson, a materials science professor at Caltech in Pasadena, asks him "Try to tear it." Lemley first pulls it gently … then with all his might with no results. "See if you can cut it," Johnson's postgraduate assistant Jason Kang says to Lemley handing him a one inch long, quarter inch wide, thinner than a dime piece of the same metal. Lemley squeezes down as hard as he can with heavy-duty wire cutters but the metal will not cut. He tries again with both hands still nothing. A steel sphere dropped on a rigid plate of the stuff bounces like a rubber Super Ball.

Conventional metal would dent as the crystals that compose them were dislocated by the impact. But the plate -- an alloy made of zirconium, titanium, nickel, copper, and beryllium -- has no crystals. Any atoms that are displaced quickly snap all the way back enabling the sphere to continue bouncing with little loss of energy. To see a fully updated description and current applications of present day technology that is in use right now every day, please see Liquidmetal Technologies where they state (and I quote): "Another unique property of Liquidmetal alloys is the superior elastic limit; i.e., the ability to retain its original shape (memory) after undergoing very high loads and stress."

They go on to say: "Liquidmetal alloys possess an 'amorphous' atomic structure which is truly unique. By contrast to the crystalline structure (of conventional metals), no discernable patterns exist in the atomic structure of the unique Liquidmetal alloys. As such, properties superior to the limits of conventional metals can be achieved."

In the end, La Paz concludes that the object that went down that night was an unoccupied probe from another planet. The bio-searcher leans toward a feeling that the incident is possibly connected somehow to the Trinity Site or maybe even Robert H. Goddard (the father of American rocketry).

In 1930, Goddard moved to Roswell. By December 30th of that year, one of his early test rockets had reached a height of 2,000 feet and a top speed of 500 mph. By May 1935, a flight reached a record height of 7,500 feet. By the 1940s, Goddard was recruited for the War effort by the U.S. Navy and lured away from Roswell leaving his launch facility for the last time on July 4, 1942. His initial program in a roundabout and somewhat disconnected way with the arrival of German scientists morphed into the V-2 program headquartered in White Sands. Goddard died August 10, 1945.

On July 16, 1945 in a location somewhat toward the west not far from Roswell, at 5:29:45 a.m. Mountain War Time at the Trinity Site, White Sands, New Mexico, the first nuclear explosive device on Earth was set off in turn creating a huge brilliant flash of light on the leading edge of the Earth's AM terminator about 12 days after her orbital aphelion. The Roswell Incident occurred almost 2 years to the date later. That would mean the expanding light bubble would have crossed the one light-year threshold one year earlier. Taking into consideration such formulas as the Drake Equation, to our knowledge nothing significant lies within that threshold.

Taken together, even though on Monday morning June 30, 1947 the bio-searcher observed 2 discs fly over his truck at a high rate of speed somewhere outside Williams, Arizona (see) and has witnessed the memory foil personally (even having been in possession of some on his own as cited above), he is not totally convinced the object that went down is extraterrestrial (but without indicating if it wasn't except for the possibility of a "time warp", then where did it come from -- his theories paralleling but predating Wernher von Braun's mentor, mathematician and physicist Dr. Hermann Oberth quoted as saying the Roswell craft should not be considered a spacecraft but a time machine).

The bio-searcher speculates at the most, the object was probably a Sputnik-type thing (an orbiting satellite) which some 10 years before in 1947 was something nobody knew anything about and probably why nobody could figure out what it was --- rather than an entry probe of some type (no doubt more Pioneer 10 like than the clunky and primitive 1957 Earth attempt Sputnik 1 that the bio-searcher was familiar with at the time and referring to).

More than likely its orbit began decaying. It started dragging the upper atmosphere, then broke apart scattering the lightweight and apparently unburnable material all over the Brazel ranch (see). The much larger and heavier vehicle (possibly no more than a shuttle craft, some sort of entry vehicle, or remote control drone search unit) ended up on the side of the Capitan Mountains, going down for unknown reasons in an effort to locate or retrieve the downed satellite and the reason for so much radar activity over the central New Mexico area in the days preceding the crash.

The bio-searcher suggests it was quite possible the craft may have landed previously and maybe even on more than one occasion to pick up or obtain the most important material or parts. Thus the reason why only so much of nothing but scraps and pieces remained at the debris field.

In 1960 a professor by the name of Ronald Bracewell presented a solution to the extraterrestrial probe aspect and "origin" differences. He proposed that advanced civilizations beyond the Solar System could make and send out automated spacecraft hoping to make contact with other civilizations. The probes (which have since come to be called Bracewell Probes) would be designed to be self-sufficient with a high degree of artificial intelligence.

Upon arrival in a new star system, the probe would place itself into orbit within the star's habitable zone and scan for radio transmissions. If radio signals were discovered, the probe would be able to communicate with a delay of no more than a few minutes (or possibly even seconds) rather than the need to communicate across the vast distances of interstellar space. Data stored aboard the probe could be easily transmitted in exchange for data from the resident civilization. Information obtained by the visiting probe would then be retransmitted back across the interstellar space to the parent system.

Such an object -- now catalogued as 1991 VG -- made an ultra-close approach to the Earth in December, 1991 following in a path of a near Earth-like orbit. Rapid brightness fluctuations and size solidly point toward the object being an artificial body rather than an asteroid.

In a weak moment one day, the bio-searcher did imply the possibilities of extraterrestrials with an off-the-cuff remark. Even though the timing seems wrong because of the amount of days between the downed craft stumbled across by the archaeologist on the morning of Saturday July 5, 1947 and the bio-searcher sighting of 2 discs almost a week before on Monday morning June 30, he did however, mention that he sometimes wondered if the objects that went over his truck outside of Williams, Arizona that morning may not have actually been flying in tandem as it appeared but instead one chasing the other with one just trying to get away.

He said that we just are assuming they are allies or somehow know each other in a friendly fashion. But they could have been mortal enemies in some sort of fight to the death, in the end leaving one disc destroyed and scattered all over the desert floor. When the existence of Bracewell Probes and remote control drones or search units were interjected into the equation, the bio-searcher questioned the need for 2 such units unless one was a command craft and the other a mule or scavenger unit. As well, he did not rule out the possibility of one disc having malfunctioned, turned rogue, or being programmed in such a way that the other was just trying to stop it.

DIRECTIONS TO CRASH SITES:

The Archaeologist Site

Unless you hike in, reaching the site (which is on the north face of Capitan Mountain) requires the use of a four-wheel drive vehicle. From U.S. 285, turn west onto Pine Lodge Road. Travel 45 miles just past mile marker 33. When you see the sign for Boy Scout Mountain, turn onto Forest Road and go 3.7 miles. Take a left on Jeep Road and drive for about a mile. After crossing a creek and climbing the hill, you will see a stone campfire ring on the left. Take the short trail to the crash site which is a rock 5 feet tall and split in two.

The Brazel Site

The Brazel Site is on Bureau of Land Management Property which is open to the public. But getting there requires passing through a private ranch gate. Permission should be requested before entering and always ensure the gate shut.

From Corona, head east on N.M. 247 (listed as old SR 42 on map link below). Turn right at the Corona Compressor Station sign just past mile marker 17. Head 16 miles from the turnoff to the site. NOTE: The person who oversees the ranch is not partcilarly fond of having tourists trampling all over the area and has been known to specifically give wrong directions to the exact location (thought to be 105:18:20W - 33:56:40N, albeit sometimes listed as 105°18.41'W - 33°56.35'N, between Gallo Canyon and Hasperos Canyon).